Insects and Minibeasts

Jinny Johnson

D0178293

KINGFISHER

KINGFISHER

First published 2011 by Kingfisher
This edition published 2014 by Kingfisher
an imprint of Macmillan Children's Books
a division of Macmillan Publishers Limited
20 New Wharf Road, London N1 9RR
Basingstoke and Oxford
Associated companies throughout the world
www.panmacmillan.com

Illustrations by: Peter Bull Art Studio

ISBN 978-0-7534-3737-7

Copyright © Macmillan Children's Books 2011

All rights reserved. No part of this publication may be
reproduced, stored in or introduced into a retrieval system,
or transmitted, in any form or by any means (electronic,
mechanical, photocopying, recording or otherwise), without
the prior written permission of the publisher. Any person who
does any unauthorized act in relation to this publication
may be liable to criminal prosecution and civil claims for damages.

1 3 5 7 9 8 6 4 2
1TR/0114/WKT/UG/128MA

A CIP catalogue record for this book is available from
the British Library.

Printed in China

This book is sold subject to the condition that it shall not,
by way of trade or otherwise, be lent, resold, hired out,
or otherwise circulated without the publisher's prior consent
in any form of binding or cover other than that in which it
is published and without a similar condition including this
condition being imposed on the subsequent purchaser.

Picture credits

**The Publisher would like to thank the following
for permission to reproduce their material.
(t = top, b = bottom, c = centre, l = left, r = right):**

Pages 4bl Shutterstock/Hordiena; 4cr Frank Lane Picture Agency
(FLPA)/Rene Krekels/Minden; 4–5b Photolibrary/John Warburton-Lee;
5tl Photoshot/Stephen Dalton/NHPA; 5tc Photoshot/Stephen Dalton/
NHPA; 5c Shutterstock/sad; 5tr Shutterstock/Vinicius Tupinam; 5cr
Shutterstock/Audrey Snider-Bell; 5br Photolibrary/Animals Animals; 6bl
Shutterstock/SF photo; 8cl Alamy/Laszlo Podor; 8c Photolibrary/Animals
Animals; 8br Alamy/Premaphotos; 9tl FLPA /Bjorn Van Lieshout/Minden;
9tr Naturepl/Warwick Sloss; 9c FLPA/Nigel Cattlin; 9bc Shutterstock/
Four Oaks; 9br FLPA/Bob Gibbons; 10bl FLPA/Scott Leslie/Minden;
12l FLPA/Malcolm Schuyl; 12tc Photolibrary/OSF; 12tr Ardea/Alan
Weaving; 12b Shutterstock/design56; 13tl Naturepl/Owen Newman;
13c Alamy/Graphic Science; 13tr FLPA/Fritz Polking; 13cl Science Photo
Library (SPL)/Bjorn Svensson; 13b Photolibrary/age footstock; 14tl
Ardea/Steve Downer; 16l FLPA/Ingo Arndt; 16br Naturepl/Andy Sands;
17tl Photolibrary/Garden Picture Library; 17tr Photolibrary/OSF; 17bl
Shutterstock/alle; 17br Naturepl/Kim Taylor; 18bl Naturepl/David Shale;
20t SPL/Kazuyoshi Nomachi; 20cl Alamy /Petra Wegner; 20c FLPA/
Matt Cole; 20cr FLPA/Konrad Wothe/Minden; 21tc Alamy/Nigel Cattlin;
21tr SPL/Norm Thomas; 21l FLPA/Bob Gibbons; 21br FLPA/ImageBroker/
Imagebroker; 22tl FLPA/Cisca Castelijns; 24tl FLPA/B Borrell Casals;
24c Naturepl/Kim Taylor; 25tl FLPA/Jef Meul/Minden; 25ctl Photolibrary/
Mauritius; 25tc Naturepl/Michael Durham; 25cr Photoshot /NHPA/John
Brackenbury; 25bl FLPA/Nigel Cattlin; 25br Photolibrary/Imagebroker;
26bl Naturepl/Robert Thompson; 28bl Ardea/Jean Paul Ferrero; 28br
Naturepl/Nick Garbutt; 29ctl FLPA/Michael Quinton/Minden; 29ctl
Photolibrary/Imagebroker; 29tr Naturepl/Dave Bevan; 29cr FLPA/
ImageBroker/Imagebroker; 29b Photolibrary/OSF; 30tl Photolibrary/
Bios; 30tr Shutterstock /javarman; 30ctr Naturepl/Edwin Giesbers; 30cbl
Shutterstock/Kirsanov; 30cbr Shutterstock/Kirschner; 30bl Alamy/Mira;
30br Shutterstock/Marek R Swadzba; 31tr Naturepl/Premaphotos;
31ctl Shutterstock/Alex_187; 31ctr Shutterstock /Peter Wey; 31cbr
Photoshot/Stephen Dalton/NHPA; 31br Photoshot/Daniel Heuclin/NHPA.

Contents

More to explore

On some of the pages in this book, you will find coloured buttons with symbols on them. There are four different colours, and each belongs to a different topic. Choose a topic, follow its coloured buttons through the book, and you'll make some interesting discoveries of your own.

For example, on page 6 you'll find a red button, like this, next to a flower. The red buttons are about plants.

Plants

There is a page number in the button. Turn to that page (page 23) to find a red button next to a flower in a meadow. Follow all the steps through the book, and at the end of your journey you'll find out how the steps are linked, and discover even more information about this topic.

Minibeasts and us

Record breakers

Habitats

The other topics in this book are minibeasts and us, record breakers and habitats. Follow the steps and see what you can discover!

A world of minibeasts

There are more kinds of insects on Earth than any other kind of creature. They live everywhere – except in the sea – and eat nearly every type of food. As well as insects, there are other minibeasts, such as snails, worms and spiders.

grasshopper

feeler

leg

head

thorax

A grasshopper, like most insects, has six legs, two pairs of wings and a pair of feelers, called antennae, on its head. Its body has three parts – the head, thorax and abdomen.

Moths use their feelers to sense smells and tastes.

At 6cm long, the **rhinoceros beetle** is one of the biggest of all beetles. It is one of the strongest insects, too, and can lift 850 times its own weight!

The male uses its big horns in battles with rival males.

wing

abdomen

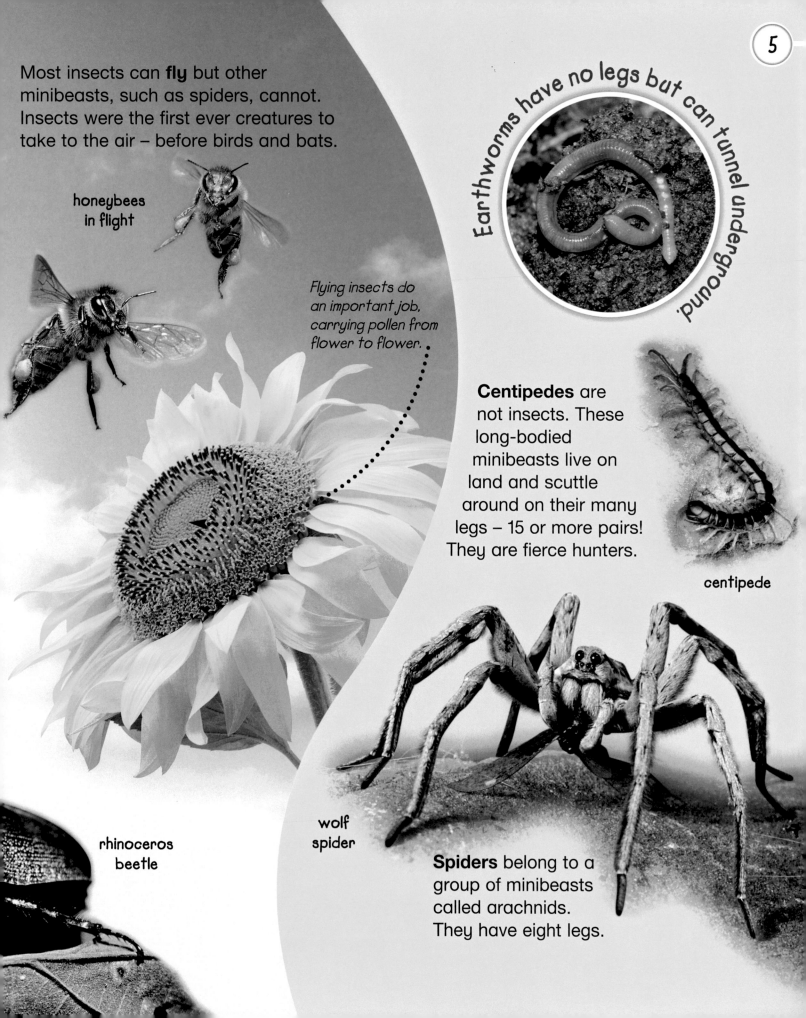

Most insects can **fly** but other minibeasts, such as spiders, cannot. Insects were the first ever creatures to take to the air – before birds and bats.

honeybees in flight

Flying insects do an important job, carrying pollen from flower to flower.

Earthworms have no legs but can tunnel underground.

Centipedes are not insects. These long-bodied minibeasts live on land and scuttle around on their many legs – 15 or more pairs! They are fierce hunters.

centipede

rhinoceros beetle

wolf spider

Spiders belong to a group of minibeasts called arachnids. They have eight legs.

From egg to adult

Most insects start life as eggs. Some insect eggs hatch into tiny versions of their parents. Butterfly eggs hatch into wriggly caterpillars. Their job is to eat and grow as much as they can before turning into beautiful butterflies.

Page 23

What is this?

① Fire ants are feasting on a monarch egg.

② Eggs are laid one at a time and stuck to leaves.

③ A newly-hatched caterpillar eats its egg.

? These are a caterpillar's stumpy little legs that allow it to crawl and climb.

Page 27

6

5

Monarch butterfly caterpillars are busy eating milkweed leaves and growing bigger and bigger. A full-grown caterpillar has made a hard outer case called a pupa or chrysalis. Inside this the caterpillar is turning into a butterfly. Nearby, a butterfly has come out of its pupa and is ready to fly away.

4 The caterpillar feeds on milkweed and grows bigger.

5 The caterpillar changes into a butterfly inside a pupa.

6 Adult monarch butterflies fly and feed on nectar.

Starting life

Most insects and minibeasts leave their eggs to hatch by themselves – their babies must start life on their own. But a few parents make sure their young have a well-stocked larder when they hatch and others even take care of their babies.

leftover nymph skin

scorpion babies on their mum's back

emperor scorpion

A shield bug guards her young.

Scorpions are surprisingly caring mums. The scorpion carries her tiny babies around on her back until they are big enough to look after themselves.

The young bugs have the same body shape as their parents.

damselfly

House fly babies are legless maggots.

Damselfly eggs hatch into water-living creatures called nymphs. When a nymph is fully grown, it crawls out of the water up a plant stem. Its skin splits and the adult damselfly struggles out, ready to fly away.

house fly

House flies lay eggs that hatch into maggots. They feed and grow, then turn into adult flies.

Baby spiders, called spiderlings, cluster together when they first hatch. Then they spin little silken threads, which act like kites to help them float away to new homes.

Dung beetles collect dung (animal poo) and roll it into a neat ball. Then, the female beetle lays her eggs in the ball. When the eggs hatch, the young have a ready meal – of yummy dung!

ball of dung

Insect nests

Honeybees make amazing nests. They live in large groups, ruled by one female, the queen. The nest is made up of lots of little six-sided cells, built with wax from the bees' bodies. Some cells contain eggs or young. Others store food.

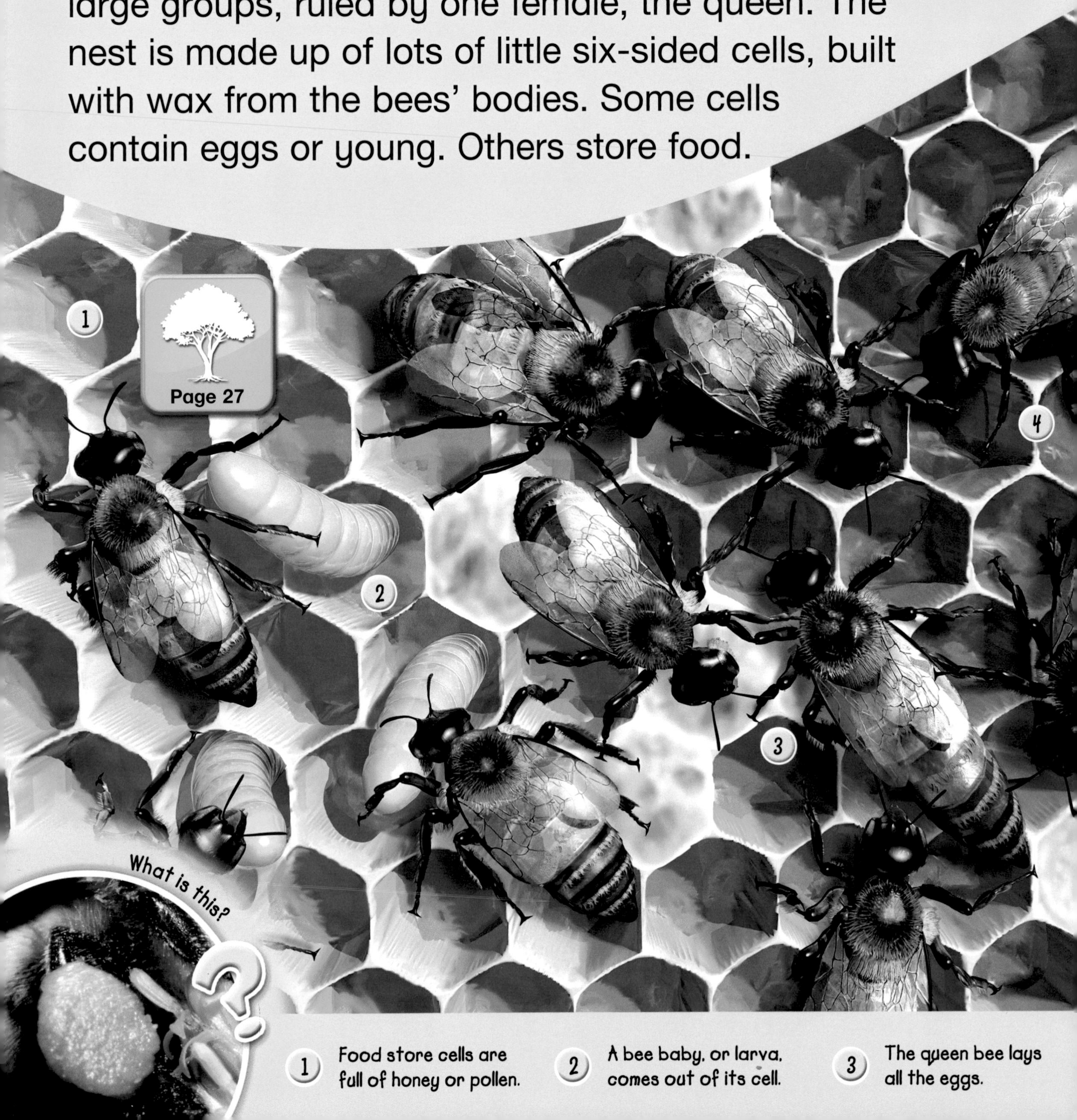

1

Page 27

4

2

3

What is this?

1 Food store cells are full of honey or pollen.

2 A bee baby, or larva, comes out of its cell.

3 The queen bee lays all the eggs.

? This is a ball of pollen. The bee carries pollen in a special, hair-lined area on each back leg.

There's always lots to do in a bees' nest. Here worker bees look after the eggs and young and keep the nest clean. Other workers gather pollen and nectar, a sweet liquid made by plants. Bees collect pollen to feed to their young and nectar for making honey to feed on in winter. Can you see the queen at the centre of the nest?

5

Page 22

Page 15

6

4 Worker bees do all the work for the colony.

5 The sealed cells contain eggs or young.

6 A worker gathers pollen and nectar from a flower.

There are tiny spiderlings inside the nursery web.

weaver ants' nest

The nursery web spider carries her eggs with her while they develop. Just before they hatch, she spins a special nursery web and puts her eggs inside. The web keeps the young safe, but she also stays on guard nearby.

Weaver ants make their nests with living leaves on trees. A group of worker ants pull the leaf edges together. Then more workers stitch the edges using silk they squeeze from their own young.

Making a nest

Many insects and minibeasts are expert nest builders. Ants and termites live in big groups and make nests from earth, leaves or bits of wood. Some spiders protect their young with sheets of silk, while potter wasps make little mud nests for each of their babies.

Termites have soft bodies and cannot live in the open for very long. They make nests on the ground or in trees to protect themselves. Ground nests can be 6m high – that's taller than three people standing on top of each other.

ground nests

The towers are made of mud and termite spit. Most of the living quarters are underground.

tree nest

Tree nests are made from pieces of wood and termite spit. The termites also make little passageways on the branch or trunk for moving to and fro under cover.

termite

Inside this pile of earth and pine needles is an ants' nest.

The potter wasp lays her egg in a pot-shaped mud nest. She then puts in some caterpillars and seals up the nest. When the baby wasp hatches, it has a well-stocked larder of food.

The wasp paralyzes the caterpillars with her sting before she puts them in the nest.

What is this?

1. an adult green darner dragonfly
2. The whirligig beetle uses its legs to paddle in circles.
3. The water boatman feeds on plants.

Page 30

Water creatures

Lots of insects and minibeasts live in and around ponds, rivers and streams. Some, like dragonflies, start their life in water and then take to the air. Others spend their whole life in the water and have legs shaped like little paddles for swimming.

On the stream bed, a fierce dragonfly nymph has just caught a young crayfish. Nearby, a giant water bug chases tadpoles. Water boatmen are not hunters. They feed on tiny water plants, which they gather with their front legs. Whirligig beetles and water striders dart around the surface catching small prey.

Page 22

Page 30

These are a dragonfly nymph's strong jaws. The lower jaw shoots out to catch prey.

Life in the water

Most water-living insects and minibeasts still need to breathe air. Some, like the diving beetle, take air down from the surface. Others have breathing tubes like built-in snorkels. And the water spider makes its own diving bell.

breathing tube

front legs for grabbing prey

Before diving to catch prey, the **great diving beetle** collects bubbles of air under its wings. These allow it to breathe while underwater. The beetle swims well using its fringed back legs like oars.

great diving beetle

The water stick insect is a kind of bug, not a stick insect at all. It hangs upside down from the water surface and breathes through a tube at the end of its body as it watches for prey.

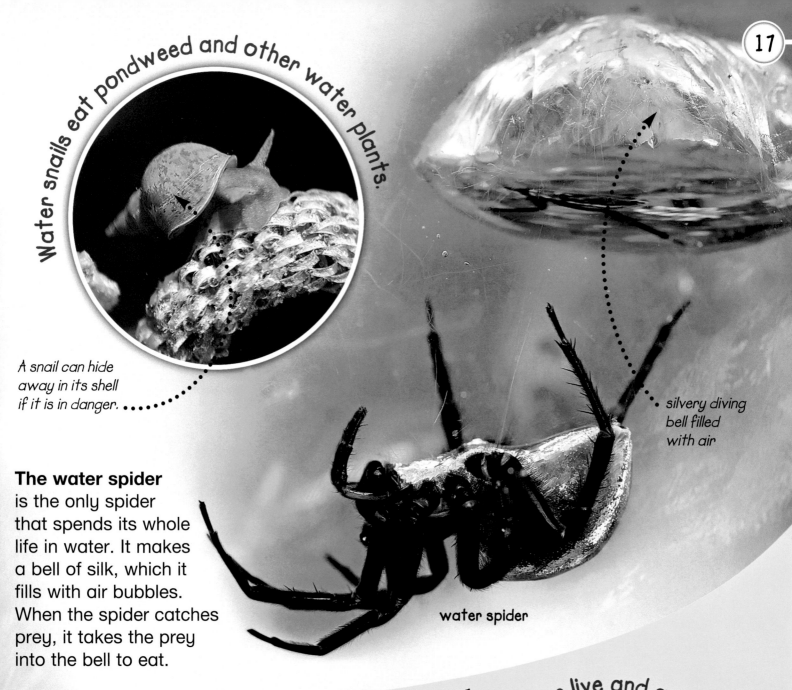

Water snails eat pondweed and other water plants.

A snail can hide away in its shell if it is in danger.

silvery diving bell filled with air

The water spider
is the only spider
that spends its whole
life in water. It makes
a bell of silk, which it
fills with air bubbles.
When the spider catches
prey, it takes the prey
into the bell to eat.

water spider

You can see an adult caddis fly on page 24 and an adult mosquito on page 25.

Caddis fly
eggs hatch into
young (larvae) like little
caterpillars. To protect its
body, the larva makes a
small case from leaves,
twigs and tiny stones,
held together with silk.

caddis fly
larva

Mosquito larvae live and grow in water.

Mini hunters

Lots of insects and minibeasts are fierce predators, just like larger animals, and life can be dangerous on the jungle floor. These little hunters have weapons such as poisonous stings or sharp claws for attacking their prey.

What is this?

1 An assassin bug slurps up its prey.

2 A scorpion uses its sting to kill prey.

3 Larger soldier ants guard the ant troop.

? This is a scorpion's sting. The scorpion swings the sting forward to inject poison into its prey.

Page 27

A troop of army ants marches across the forest floor. One group of the ants has attacked a much larger creature – a tiger beetle. A praying mantis munches prey it has seized with a lightning-fast movement of its front legs. Nearby, a tarantula and a centipede wait to catch insects running away from the ants.

Page 14

4 A praying mantis crushes its prey with strong jaws.

5 A tarantula waits for prey. It has a poisonous bite.

6 The centipede feeds on insects and spiders.

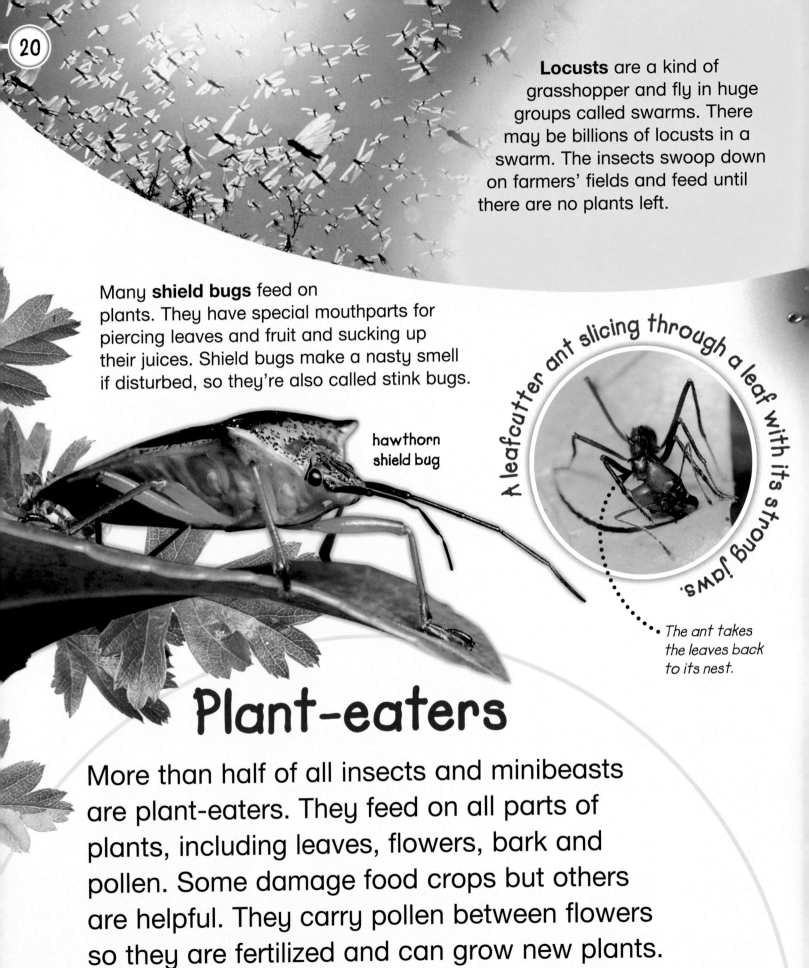

Locusts are a kind of grasshopper and fly in huge groups called swarms. There may be billions of locusts in a swarm. The insects swoop down on farmers' fields and feed until there are no plants left.

Many **shield bugs** feed on plants. They have special mouthparts for piercing leaves and fruit and sucking up their juices. Shield bugs make a nasty smell if disturbed, so they're also called stink bugs.

A leafcutter ant slicing through a leaf with its strong jaws.

hawthorn shield bug

The ant takes the leaves back to its nest.

Plant-eaters

More than half of all insects and minibeasts are plant-eaters. They feed on all parts of plants, including leaves, flowers, bark and pollen. Some damage food crops but others are helpful. They carry pollen between flowers so they are fertilized and can grow new plants.

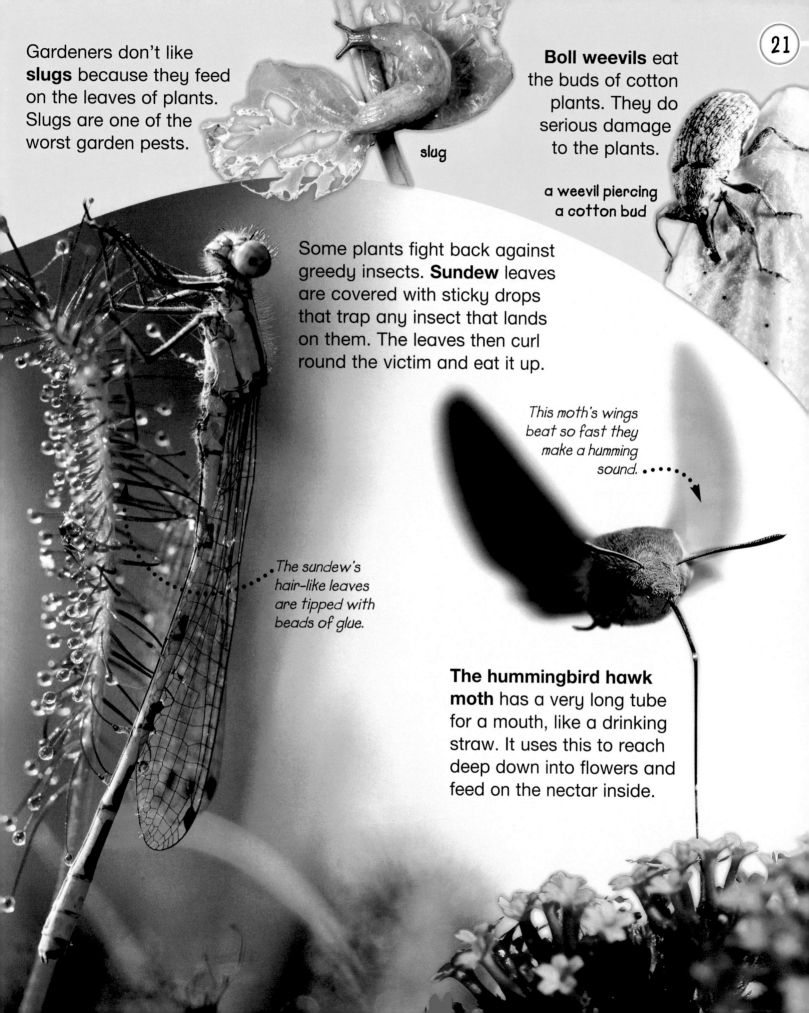

Gardeners don't like **slugs** because they feed on the leaves of plants. Slugs are one of the worst garden pests.

slug

Boll weevils eat the buds of cotton plants. They do serious damage to the plants.

a weevil piercing a cotton bud

Some plants fight back against greedy insects. **Sundew** leaves are covered with sticky drops that trap any insect that lands on them. The leaves then curl round the victim and eat it up.

This moth's wings beat so fast they make a humming sound.

The sundew's hair-like leaves are tipped with beads of glue.

The hummingbird hawk moth has a very long tube for a mouth, like a drinking straw. It uses this to reach deep down into flowers and feed on the nectar inside.

What is this?

Page 30

Page 19

1. The dragonfly catches its prey on the wing.

2. Honeybees carry pollen from flower to flower.

3. A marbled white butterfly feeds on nectar.

Fast fliers

Being able to fly gives insects a huge advantage over other minibeasts. They can move around much more quickly than land-based minibeasts. Flight allows insects to find food easily and to escape from their enemies. Most insects have two pairs of wings.

4 Hoverflies can hover and even fly backwards.

5 The cricket can jump with its long legs as well as fly.

6 A lacewing has four see-through, veined wings.

6

In a summer meadow, honeybees and hoverflies visit flowers to gather pollen and nectar, while butterflies perch on the flowers and sip the sweet nectar. Ladybirds fly from plant to plant hunting aphids. A speedy dragonfly zooms in searching for insect prey, and startles a bush cricket.

3

5

4

Page 11

These are the tiny scales on a butterfly's wing that make the pretty patterns.

The caddis fly has hairs on its body as well as its wings.

The hard front wings act as cases for the back wings.

A caddis fly looks very like a moth but its wings are covered with tiny hairs instead of scales. (See its larva on page 17.)

a ladybird in flight

Like all beetles, a **ladybird** uses its fragile back wings for flying. The rest of the time these wings are folded away, protected by the spotty front wings.

Special wings

Not all insect wings are the same. A beetle's front wings are very hard and protect the delicate back wings. Butterflies have two pairs of colourful, velvety wings, while flies have one pair of see-through ones. Wings aren't just for flying, either. Some insects use them to make sounds.

A bumblebee's wings beat at least 130 times a second.

Mosquito wings beat so fast they make a whining noise.

Male crickets chirp to attract a mate. They make their song by rubbing together their wings. A scraper on one wing rubs against a ridge on the other wing.

great green bush cricket

A bumblebee makes a loud buzzing sound as it flies around. This is not the sound of its wings beating but of its flight muscles vibrating (wobbling).

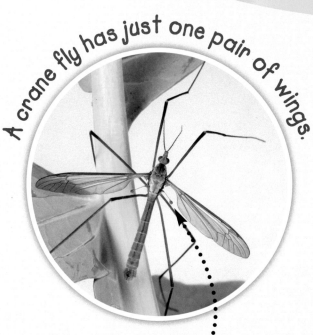

A crane fly has just one pair of wings.

Birdwing butterflies are the world's largest butterflies. The Cairns birdwing has an amazing wingspan of up to 16cm – yet it's only half the size of the record-breaking Queen Alexandra's birdwing!

Cairns birdwing

Knobbed stumps called halteres help the crane fly keep its balance.

Watch out!

Life is dangerous for insects and minibeasts. They hunt each other, and there are always birds and other creatures ready to snap them up, too. But insects and minibeasts have many ways to avoid danger and warn off enemies.

What is this?

1 A wasp can sting any attacker.

2 The stink bug's bad smell puts off predators.

3 Ants bite or sting to attack and defend themselves.

? This is the eyespot on a caterpillar's bottom. It makes the caterpillar look big and fierce.

Page 19

Page 30

Page 15

A colourful locust, a kind of grasshopper, leaps up from the ground, flashing its wings and startling any predator. A stink bug warns off its enemy with a nasty smell, while a bombardier beetle squirts boiling-hot chemicals at some attacking ants. The ants are no match for the beetle, even though they can give a painful sting.

4 The bombardier beetle sprays hot chemicals.

5 The locust can jump away from danger.

6 Foamy cuckoo spit protects a froghopper's young.

Staying safe

To avoid being spotted, some insects and minibeasts look like something else, such as a leaf or a thorn. Others are the same colour as the flowers they live among. Some stay safe by disguising themselves as other, more dangerous, creatures.

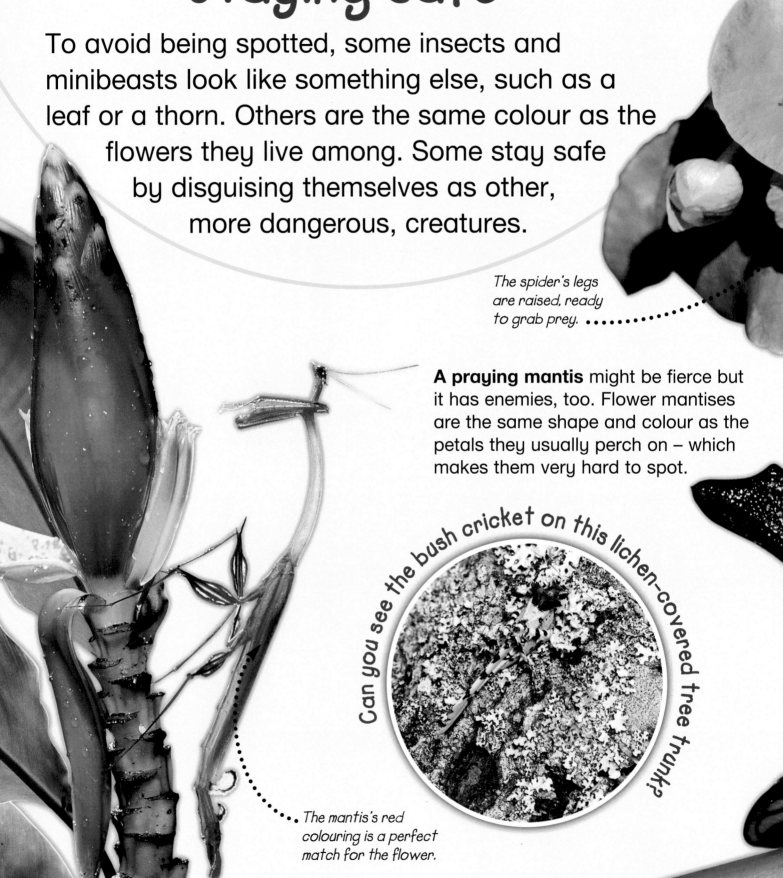

The spider's legs are raised, ready to grab prey.

A praying mantis might be fierce but it has enemies, too. Flower mantises are the same shape and colour as the petals they usually perch on – which makes them very hard to spot.

Can you see the bush cricket on this lichen-covered tree trunk?

The mantis's red colouring is a perfect match for the flower.

crab spider on
a buttercup

wasp

Some insects **disguise themselves** as more dangerous insects. The hoverfly doesn't sting, but its stripy body makes it look like a wasp or bee, which does. Thanks to this clever disguise, most predators steer clear of the hoverfly.

Crab spiders like to hide in flowers so they can ambush their prey. Some can even change colour to match the flower they are on.

hoverfly

These strange-shaped **bugs** look just like thorns on a branch. As long as they stay still, birds and other insect-eaters won't notice them.

This is the female thorn bug – she's a slightly different shape from the male.

Habitats

Honeybees live almost all over the world and in almost every **habitat**. They build their nests in woodlands, fields, gardens and tropical forests – anywhere that has flowers they can visit for food.

wild honeybee nest

Africa has hot, dry **grasslands** called savannah, where few large trees grow. The savannah is home to billions of insects, including termites, ants, locusts and grasshoppers.

Plants

Milkweed plants contain poisonous, milky sap. Monarch caterpillars can eat the leaves without being harmed. They take in the poison and this makes them taste nasty to predators.

Many flowers contain **nectar,** a sweet liquid that insects like to eat. Insects come to the plant to feed on nectar and bees use nectar for making honey.

butterfly on a cistus flower

Minibeasts and us

bee-keeper

Honeybees are the most useful insects. We love the honey they make for their food stores and some people also eat pollen. Beeswax is used to make furniture polish and some candles.

bee on apple blossom

There are **fewer and fewer bees** and this could be a problem for farmers. Honeybees pollinate apples, carrots and other food crops. Scientists think that pollution and pesticides might be harming bees.

Record breakers

Monarch butterflies make the **longest journeys** of any insect. They spend the winter in southern California or Mexico. In spring, they fly as far as 4,000km north to find the plants their caterpillars like to eat.

Froghoppers or spittlebugs are **champion jumpers,** able to jump 100 times their own length. If you could do that, you could leap over your school!

froghopper

More to explore

praying mantis

Tropical rainforest grows in parts of the world where it is hot and wet all year round. There are more kinds of insect and minibeast in rainforests than any other kind of animal.

Like all animals, insects need **water.** Most need water to drink. Some, such as mosquitoes and dragonflies, lay their eggs in water. The eggs hatch into swimming larvae that spend the first part of their lives in water.

mosquito eggs

Pollen looks like dust in the centre of a flower. When an insect feeds on nectar, its body picks up pollen. Later, the pollen may brush off on another flower of the same kind, allowing that plant to produce seeds.

Water plants can live in water or floating on the surface. They provide places for insects and minibeasts to shelter and they also shade the water.

water lily

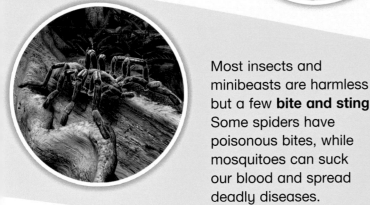

Most insects and minibeasts are harmless but a few **bite and sting**. Some spiders have poisonous bites, while mosquitoes can suck our blood and spread deadly diseases.

Some insects are pests and **damage plants and crops.** Locusts are grasshoppers that sometimes form swarms. Hungry locusts can strip a farmer's fields in a day.

desert locust

Dragonflies are **speedy** as well as beautiful. Large dragonflies can fly at an amazing 55km/h – that's faster than a car drives in the city.

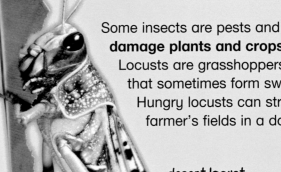

Butterflies and moths have the **biggest wings** in the insect world. The atlas moth's wings are a whopping 30cm across.

atlas moth

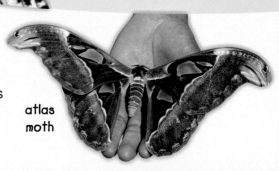

Index